Look and Learn

A First Book about
Animal Homes

For a free color catalog describing Gareth Stevens Publishing's list of high-quality books and multimedia programs, call 1-800-542-2595 (USA) or 1-800-461-9120 (Canada). Gareth Stevens Publishing's Fax: (414) 225-0377.

Library of Congress Cataloging-in-Publication Data

Tuxworth, Nicola.
　A first book about animal homes / by Nicola Tuxworth.
　　　p. cm. — (Look and learn)
　Includes bibliographical references and index.
　Summary: Photographs and simple text describe where such animals as chickens, cows, rabbits, fish, bees, hamsters, and dogs are kept.
　　ISBN 0-8368-2285-4 (lib. bdg.)
　1. Animal housing—Juvenile literature. [1. Animals—Habitations.] I. Title. II. Series: Tuxworth, Nicola. Look and learn.
SF91.T89　1999
636.08'31—dc21　　　　　　　　　　　　98-31779

This North American edition first published in 1999 by
Gareth Stevens Publishing
1555 North RiverCenter Drive, Suite 201
Milwaukee, WI 53212 USA

Original edition © 1997 by Anness Publishing Limited.
First published in 1997 by Lorenz Books, an imprint of Anness Publishing Inc., New York, New York.
This U.S. edition © 1999 by Gareth Stevens, Inc.
Additional end matter © 1999 by Gareth Stevens, Inc.

Senior editor: Sue Grabham
Stylists: Marion Elliot and Isolde Sommerfeldt
Special photography: Lucy Tizard
Design and typesetting: Michael Leaman Design Partnership

Picture credits: Ecoscene 7bl, 21tr and br. Holt Studios International 6tr, 7tl, 9b. Planet Earth Pictures 20bl.

All rights reserved. No part of this book may be reproduced, stored in a retrieval system, or transmitted in any form or by any means, electronic, mechanical, photocopying, recording, or otherwise without the prior written permission of the copyright holder.

Printed in Mexico

1 2 3 4 5 6 7 8 9 03 02 01 00 99

Look and Learn

A First Book about

Animal Homes

Nicola Tuxworth

Gareth Stevens Publishing
MILWAUKEE

Stable

Some horses sleep in a cozy stable at night.

Horses often eat grass outside in the fields.

Inside the stable, there is food and water for the horse.

hay

feed

A horse can see over its stable door.

water

Chicken coop

Chickens eat and sleep in a chicken coop.

Chickens often lay eggs in the coop.

chick

The coop has a door and a ramp.

Chickens eat corn.

At night, the coop keeps the chickens safe from hungry foxes.

Cow barn

Inside a cow barn, there is hay to eat and a soft, straw bed.

calf

Cows look after their calves in the cow barn.

Cows live together in big groups called herds.

Rabbit hutch

Pet rabbits live in a wooden cage. It is called a hutch.

The hutch has a bedroom for the rabbit.

Rabbits need special food.

Rabbits like to nibble on crunchy carrots and lettuce.

They sip water from a bottle.

Aquarium

Pet fish live in an aquarium. A pump keeps the water fresh.

Can you see one fish or two fish?

Fish poke around in the gravel for tasty tidbits.

gravel

fish food

Beehive

Bees are kept in wooden boxes called hives.

The beekeeper wears special clothes.

Honeybees collect pollen from flowers.

Thousands of bees live in the hive.

Bees make honey inside the hive.

honeycomb

Hamster cage

A hamster can run around in its cage.

Sawdust makes a cozy bed.

Hamsters run up and down ladders ...

and hide in little houses.

hamster food

Kennel

Some dogs sleep outside in a warm, dry kennel.

Dogs like to play with toys.

Some dogs sleep indoors in a special basket.

Dogs need to be taken for walks.

Nesting box

Some birds build their nests in a special wooden nesting box.

Sheep's wool makes a soft nest lining.

How many speckled eggs has this little bird laid?

Can you see the food on this bird feeder?

Birds like to eat nuts.

Can you remember where these animals live?

Glossary/Index

coop: a building where chickens are protected. (pp. 6-7)

herd: several similar animals, like cows, that live together. (p. 9)

hutch: a cage in which a pet rabbit lives. (p. 10)

kennel: a crate or small house in which dogs sleep and are protected. (p. 18)

nesting box: a birdhouse; a small, wooden box where birds lay their eggs. (p. 20)

pollen: the tiny parts of a flower that help make new seeds. (p. 15)

speckled: covered with spots. (p. 21)

stable: a building or barn that shelters horses and other animals. (pp. 4-5)

More Books to Read

Animal Habitats: The Best Home of All. Nancy Pemberton (Child's World)

Animal Homes. Brian Wildsmith (OUP)

Real Baby Animals (series). Gisela Buck & Siegfried Buck (Gareth Stevens)

Who Lives Here? Maggie Silver (Sierra)

Videos

Animal Homes. (Coronet, Multimedia)

A Visit to the Farm. (New World Video)

Web Sites

www.colapublib.org/children/kids/wild/animal/homes/index.html

www.davisfarmland.com/

Some web sites stay current longer than others. For further web sites, use your search engines to locate the following topics: *aquariums, beehives, birdhouses, chickens, cows, dogs, habitats, hamsters, horses,* and *rabbits*.